A BIORENAISSANCE

The human place in nature –
past, present and future

STEPHEN BOYDEN

ACKNOWLEDGEMENTS

This booklet would never have been completed were it not for the encouragement and support of my granddaughter, Karina Bontes Forward. She has devoted countless hours to reading, editing and proofreading the manuscript and to talking with me about the text. Thank you, Karina.

My sincere thanks go to Leo Robba and Matthew Lahoud for all their work on the artwork and design of the booklet and general preparation for publication, and to Adrienne Richards for the picture on the cover. Also to Jane Olsen for proofreading, to Luis Ramirez and Jorge Bontes for preparing figures, to Jane Milburn for some very useful advice, and to Sally Tsoutas for her photographs.

I would also like to express my heartfelt thanks to the students and colleagues in human ecology who contributed so much to my thinking and enjoyment of life over so many years.

Last, but by no means least, very special thanks to my wonderful family who, for so long, have tolerated my idiosyncracies, and whose love and support have kept me going.

COVER ARTIST STATEMENT

Adrienne Richards
On Earth, as it will be in the Heavens, 2023
glazed stoneware and porcelain

My work is inspired by Stephen Boyden's books about Biohistory. The title alludes to humans' capacity for culture, story-telling and our voracious appetite for exploration. The medium is glazed ceramic, where the Earth (clay) has been transformed, a manifestation of human discovery and control of fire. The shards are fossil-like remains from the natural world, and symbols of human activity.

Editor: Karina Bontes Forward
Art Direction & Design: Dr. Leo Robba, Matthew Lahoud

Old Hillside Publishing
ISBN: 978-0-6459295-0-8
Copyright © Stephen Boyden 2023
biorenaissance.net

CONTENTS

PROLOGUE	5
PART 1: A WAY FORWARD	7
A wave of new understanding	8
A biosensitive society	10
Biocentres	13
PART 2: THE BIONARRATIVE IN A NUTSHELL	19
The first 4000 million years	20
Photosynthesis	24
Reproduction	25
The Earth's recycling system	26
Extinctions	27
Homo sapiens appears on the scene	28
Population and energy	32
Greenhouse gases	34
Deforestation	35
Farming	35
Waste production and pollution	36
Climate change and the oceans	38
Spacecraft	38
Loss of biodiversity	38
Environmentalism and the green movement	38
PART 3: SOME BIOHISTORICAL PERSPECTIVES	40
Human culture: a new kind of force in the biosphere	41
Health and disease	43
Religion	48
Warfare	50
Vegetarianism and veganism	54
Biometabolism and technometabolism	56
Technoaddiction	58
A Generational Perspective	58
Human rights and the rights of nature	59
Pessimism and optimism	60
PART 4: PERSONAL PERSPECTIVES	62
Nature	64
Evolution	65
Ethics	67
Spirituality	68

PROLOGUE

This booklet is an outcome of my work in biohistory at the Australian National University from 1965 to 1990, and afterwards during my retirement.[1]

An important feature of the biohistorical approach is that it recognises the emergence of human culture as a new kind of force in biological systems. Through its influence on human behaviour, culture has impacts not only on humans themselves, but also on the rest of the living world.

This is not an academic treatise. It makes no attempt to review the literature in this field, or to discuss the work of others. It is simply a summary of my own personal thoughts that I would like to communicate at this time in my life, plus some general discussion on biohistorical themes.

I am among those who appreciate that human activities are now on a scale and of a kind that threaten the survival of civilisation, and perhaps of the human species. The only hope for the future lies in a radical transformation across the cultures of the world, leading to big changes in human activities, and to societies that are truly in tune with, and respectful of, the processes of life that underpin our existence.

We can call this transformation a 'biorenaissance' because human societies in the past have been ecologically sustainable, and human cultures have held deep respect for the natural world.[2]

This booklet consists of four parts.

The first part summarises my thoughts about the human situation on Planet Earth today, and a possible way forward to a healthy and ecologically sustainable society of the future. I put the view that the survival of civilisation will depend on a wave of new understanding sweeping across the cultures of the world – understanding of the story of life on Earth and the human place in nature. I refer to this story as the Bionarrative.

---

1   Biohistory has been defined as the study of human situations, past and present, against the background of, and as part of, the story of life on Earth (see Boyden, S. 1987. *Western civilization in biological perspective: patterns in biohistory*. Oxford University Press, Oxford and Boyden, S. 1992. *Biohistory: the interplay between human society and the biosphere - past and present*. Parthenon/UNESCO, Paris).
2   From *bios*, ancient Greek, meaning 'life', and *renaissance*, French, meaning 'rebirth' or 'a revival of renewed interest in' (Oxford Languages).

The second part is an extremely short version of the Bionarrative.

In **Part 3** I discuss some key issues in biohistorical perspective.

Part 4 summarises my personal responses to learning about life on Earth.

The transition to a society that is in harmony with nature will involve a sequence of cultural and social steps. To facilitate thinking and communicating about the transition, I have found it necessary to introduce several new terms, which are defined in **Table 1**.

TABLE 1: DEFINITIONS

Biohistory	The study of human situations against the background of, and as part of, the story of life on Earth.
Biorenaissance	The cultural and social transformation to a society that is in harmony with nature.
Bionarrative	The story of life on Earth, including the activities and impacts of *Homo sapiens* in recent times.
Biounderstanding	Understanding the story of life on Earth and the human place in nature.
Biosensitive	In tune with, and respectful of, the processes of life.
Bioreverence	A worldview that is based on biounderstanding and that: • holds profound respect for the processes of life • perceives the achievement of harmony with nature as supremely important, to be given the highest priority in human affairs • embraces a vision of a society of the future that is in tune with and respectful of the processes of life.

PART 1:
A WAY FORWARD

A WAVE OF NEW UNDERSTANDING

Homo sapiens has been in existence for around 300,000 years. The last 200 years has seen massive growth in the human population and per capita resource and energy use and waste production, with ever-increasing impacts on the ecosystems of our planet.

Apart from weapons of mass destruction, climate change is at present the most critical threat; but there are other anthropogenic threats to the sustainability of the living systems on which we depend (see Pages 32-38). If present conditions continue unabated, the collapse of civilisation is inevitable.

The survival of civilisation will require a shift to a different kind of society – a society in which human activities are in harmony with the processes of life within and around us. I refer to such a society as a *'Biosensitive society'*.[3]

Unfortunately, the prevailing cultures across the world today are simply not attuned to these ecological realities. They have lost sight of the fact that we humans are completely dependent on healthy ecosystems for our survival and wellbeing, and they have no grasp of the magnitude and gravity of current human impacts on the biosphere. They are driving us towards ecological catastrophe.[4] There will be no effective transition to ecological sustainability unless there comes about a radical transformation in the worldviews and priorities of these cultures.

In my view, the only hope for the survival of civilisation lies in the possibility of a great wave of new understanding sweeping across the cultures of the world – understanding of the story of life on Earth and the human place in nature. I refer to this story as the *'Bionarrative'*, and I call this kind of understanding *'Biounderstanding'*.

---

3 The term 'biosensitive' is introduced because there is a need for a single word to describe a society with these characteristics. The expression 'ecologically sustainable' is widely used. Of course, society must be ecologically sustainable, otherwise in the long term it cannot continue to exist. However, ecological sustainability is surely the bottom line. We must aim for a society that positively promotes health and wellbeing in the ecosystems of the biosphere as well as in all sections of the human population.

4 It is true that the Green Movement has gained momentum over the past 50 years, but it still has a very long way to go (see page 38).

The Bionarrative is crucially important in two ways. First, it reminds us that we are living beings, products of several billion years of biological evolution, and that we are part of nature, and totally dependent on other living organisms for our survival and wellbeing.

Second, it alerts us to the gravity of the current ecological crisis, and to the urgent need for big changes in the scale and kind of human activities globally.

Shared understanding of the Bionarrative would lead to a worldview that holds profound respect for nature, and that perceives the achievement of harmony with the processes of life as supremely important, to be given the highest priority in human affairs.

If this worldview were embedded at the core of the dominant cultures worldwide, the prospects for the future of humanity and the biosphere would be much brighter. There is therefore an urgent need for a global movement promoting this kind of understanding.

This worldview needs a name. It encompasses environmentalism, but it has a broader meaning, because it is about life in its entirety, past and present, non-human and human, not just 'the environment'. I have wasted countless hours trying to think of a suitable term, and I have finally decided to call it Bioreverence. I am not very happy with this word, but I cannot think of anything better; so I will use it for the time being, until someone comes up with something better.[5]

If Bioreverence were embedded at the core of the dominant cultures worldwide, the future prospects for humanity and the biosphere would be much brighter. There is therefore an urgent need for a global movement promoting biounderstanding.

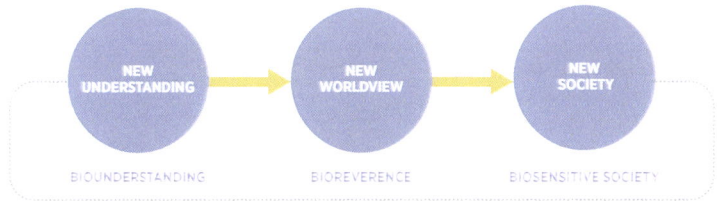

Figure 1. The Biorenaissance

---

[5] I considered the term 'biophilia', which has been used by E.O Wilson, but then decided it was not appropriate, because it has been defined as 'an innate and genetically determined affinity of human beings with the natural world.' Bioreverence is not necessarily innate. It can come about through learning.

A BIOSENSITIVE SOCIETY

The most essential feature of a biosensitive society will be a prevailing culture that views the human situation in biological perspective. Harmony with nature will be given the highest priority in human affairs.

It will be a society that is in tune with and respectful of the life processes that underpin our existence.

A biosensitive society will promote health in all sections of the human population, as well as in the ecosystems of the biosphere (**Figure 2**).

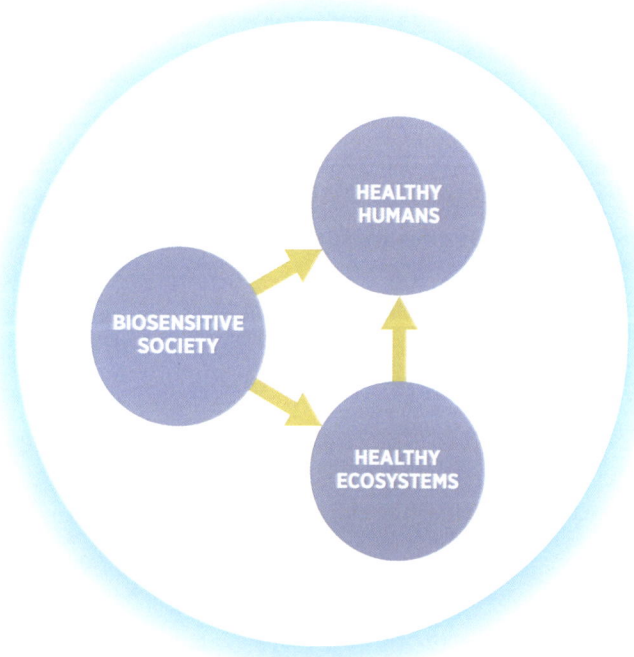

Figure 2. Biosensitivity triangle

Table 2 lists some key features of a society on the path to biosensitivity. The achievement of these objectives will require strong and enlightened government action.

Governments will also oversee a transition to a biosensitive economic system that satisfies the needs of all sections of the human population without resulting in ever-increasing consumption of natural resources and production of wastes.

TABLE 2: TOWARDS BIOSENSITIVITY – ESSENTIALS
Clean energy sources, no use of fossil fuels
Carbon sequestration through massive reforestation and other means
Ecologically sustainable human population
Promotion of healthy soils, and organic waste returned to farmland
More food produced locally
Widespread protection of biodiversity on land and in the oceans
Reduced overall use of resources and energy
Non-existence of weapons of mass destruction
No release of waste plastics into the environment
No release of persistent organic pollutants
Clean air in urban environments
Clean water supplies

At the level of individuals and families, biosensitivity will be associated with a high quality of life. Lifestyles will satisfy people's biologically determined health needs. A working list of important human health needs is presented in **Table 3** (next page).

In a biosensitive society, these health needs will be satisfied in ways that do not result in continual growth in the use of resources and energy, increasing pollution of the natural environment or loss of biodiversity. There will be more emphasis than at present on such activities as growing food, enjoying and caring for the natural environment, local sport, making music, dancing, art, theatre, cycling, and convivial social interaction. Rampant consumerism and travel powered by fossil fuels will not be features of a biosensitive society.

Attention is drawn to the psychosocial category of human health needs. Although somewhat difficult to define and measure, these intangible needs, such as a sense of purpose and the experience of conviviality, are as important for wellbeing as are a healthy diet and clean air.

TABLE 3: HUMAN HEALTH NEEDS

Physical needs	Psychosocial needs
A natural diet (i.e. a calorie intake neither less than, nor in excess of, metabolic requirements; the full range of nutritional requirements and fibre; and free of harmful contaminants and additives)	Life conditions conducive to a sense of purpose, sense of belonging, sense of challenge, sense of comradeship and love
Clean air (not contaminated with hydrocarbons, sulphur oxides, lead etc.)	An emotional support network providing a framework for care-giving and care-receiving behaviour
Clean water (free of contamination with harmful chemicals or pathogenic micro-organisms)	The experience of conviviality
Absence of harmful levels of electromagnetic radiation	Opportunities and incentives for co-operative small-group interaction
Minimal contact with parasites and pathogens, but natural contact with normal environmental microbes	Opportunities and incentives for creative behaviour
Adequate safe shelter from extremes of weather	Variety in daily experience
Noise levels within the natural range	Contact with the natural environment
Plenty of physical exercise	

BIOCENTRES

This booklet champions the view that a great wave of biounderstanding, sweeping across the cultures of the world, is a precondition for the survival of civilisation.

How might this wave of new understanding be brought about? In my view, the best hope lies in the creation of a worldwide network of public institutions of a new kind, which are dedicated to spreading understanding of life, and to promoting a vision of a healthy and ecologically sustainable society of the future. Let us call these institutions *'biocentres'*. Apart from their educational role, they would provide people with the opportunity to gather together and celebrate life on Earth in different ways.

Like religious and political institutions, biocentres can be envisaged as existing at different levels in society, from small local groups, which we can call community biocentres, to national bodies on a par with national museums and places of worship. These physical centres would be supported by an online educational network, focusing on digital and interactive elements of the biocentres and helping people to connect.

This is a global issue of outstanding importance, so we might hope that the United Nations, or one of its agencies, would play a leading role in establishing the biocentre movement. It would be completely consistent, for example, with the mission and objectives of UNESCO.

Meanwhile, there is positive action that can be taken in the short term at a national level, here in Australia, as indicated in the following proposal.

PROPOSAL FOR AN AUSTRALIAN NATIONAL BIOCENTRE

For healthy people on a healthy planet

This proposal is based on the appreciation that there is a need for a new kind of public institution in our society. There is a need for an institution that spreads understanding of life on Earth and the human place in nature, and that promotes a vision of a future society that is in harmony with nature.

Canberra, like other urban centres, has a big range of public institutions. They include a National Gallery, a Portrait Gallery, a National Library, a War Memorial, a National Museum of Australia, a National Institute of Sport, and a range of churches, mosques, temples, theatres, arts centres, and concert halls. There are also institutions focusing on non-human forms of life, like Nature Reserves, the Arboretum, the National Botanic Gardens and the National Zoo.

Yet, strangely, there is no public institution that is concerned with the story of life in its entirety, including *Homo sapiens*. This is curious, because we humans are living organisms, products of biological evolution, part of nature, and entirely dependent on the processes of life, within and around us, for our survival and wellbeing. Indeed, our civilisation is a by-product of biological evolution, also completely dependent on biological processes for its very existence.

It is therefore proposed that an Australian National Biocentre (ANB) be established in Canberra.

ANB will be for people who are interested in life on Earth and who care about the future of humankind and our planet. It will be based in one or more buildings with indoor displays, situated on a few acres of parkland, which will also have outdoor displays.

Aims

The aims of ANB will be:

- to spread understanding, across the community, of life on Earth and the human place in nature; and
- to promote a vision of a healthy and ecologically sustainable society of the future that is in harmony with nature.

Activities

Educational activities

This aspect of ANB will be modelled to some extent on the Australian Museum in Sydney, with its exhibitions, lectures, courses, publications and a website, except that the overriding theme will be different.

Themes are likely to include, for example:

The evolution of life on Earth
Biodiversity
Photosynthesis
Four ecological phases of human history
Human ecology in Australia over 50k years
Indigenous perspectives
Feral animals in Australia
Human population – perspectives
History of energy use by humans
Climate change
Clean energy
Hominid evolution
Human culture as a force in nature
Forests and humans
Microbes and nutrient cycles
Insects and humans
Human health needs – physical and psychosocial
The reproductive system
Infectious diseases and pandemics
Biosensitive lifestyles
Economics and ecological sustainability
Food production, past present and future
Warfare and weaponry in biohistorical perspective

A team of science writers will be responsible for producing the educational material.

Local universities and research institutions will make a major contribution to the activities of ANB. Community groups, NGOs, commercial organisations, and government agencies will be encouraged to mount displays consistent with the aims and philosophy of ANB.

Exchange of ideas

ANB will arrange informed dialogue among leaders from different walks of life on the social changes necessary to achieve ecological and social sustainability, and on how they might be brought about. The outcome of these deliberations will be communicated online and directly to educational and governmental authorities.

Network of Community Biocentre Groups

The ANB will establish a network of community Biocentre Groups across the nation. They will be linked with ANB and with each other.

The Biocentre Groups will provide a framework for concerned and interested people to gather together to;

- learn about life on Earth and the human place in nature;
- exchange ideas about the way forward to a healthy and ecologically sustainable society of the future;
- engage in activities consistent with the philosophy and aims of ANB; and
- celebrate life on Earth.

Social activities

Canberrans who are interested in ANB will be able to join the Friends of the Biocentre. This will provide them with opportunities to participate in programs of talks, workshops and social events, and to celebrate life on Earth through art, music, gardening and other activities.

Conclusion

The Australian National Biocentre will be a great place to visit. It could make a significant contribution to our society's transition to ecological sustainability, and it might well become a prototype for similar institutions in other places across the globe.

HISTORICAL BACKGROUND TO THIS PROPOSAL[6]

The current proposal for the establishment of an Australian National Biocentre is, in essence, an updated version of an earlier proposal. In 1965 a group of scientists in the ACT proposed to the Federal Government that there should be a new kind of public institution in Canberra which they called a Biological Centre.[7]

The proposal was based on the view that the prevailing culture at the time suffered from a serious deficiency. It had lost sight of the fact that we are living organisms, part of nature, products of biological evolution, and totally dependent on the processes of life within us and around us, for our health, wellbeing and very existence. This reality was simply not reflected in the prevailing worldview and institutional structure of society, much to the disadvantage of humans and of the living systems on which we depend.

The group, of which I was a part, argued that there was a pressing need for a new kind of public institution to help counter this cultural void, a new kind of institution to stand alongside other public institutions like art galleries, museums, war memorials, botanic gardens, and cathedrals.

The Biological Centre would be about life – its history, how it all works, and how we humans emerged through the processes of biological evolution. It would be about our own biology and about the impacts that our species is having on the rest of the living world. It would encourage dialogue on the meaning of this understanding for individuals and families and for society. There would be much more community involvement than in a conventional museum or zoo.

The proposal was strongly supported by some overseas scientists, including Julian Huxley, Konrad Lorenz and Nikko Tinbergen; and it was supported by every primary and secondary school in Canberra.

---

6   *For more detail, see* Libby Robin with Stephen Boyden. 2018. Telling the Bionarrative: a Museum of Environmental Ideas. *Historical Records of Australian Science* 29(2) 138 - 152 20 2018.

7   The group of scientists behind this proposal included R. E. Barwick, E.C.F. Bird, S. V. Boyden (Convener), J. H. Calaby, R. Carrick, D. G . Catcheside, A. B. Costin, M. F. Day, A.H. Ennor, F. J. Fenner, O. H. Frankel, H. J. Frith, S. B. Furnass, E. H. Hipsley, I. M. Mackerras, W. L. Nicholas, M. Oliphant, L. Pryor, F. N. Ratcliffe, R. Slatyer, J. D. Smyth, D. F. Waterhouse, W. K. Whitten.

An article describing the proposal was published in the International Zoo Yearbook in 1969.

The proposal was formally put to the Prime Minister, Sir Robert Menzies, in May 1965. It took the government seven years to make up its mind, and when it did so, it was positive. Canberra would have a Biological Centre.

However, just at that time, in 1972, a federal election was called, and a new government came into power; and the new Minister for the Environment, Moss Cass, reversed the decision to establish the Biological Centre.

So, there is no Biological Centre in the national capital today.

In later years, after his retirement from politics, Cass regretted his decision. In fact, for a while he was chairman of a committee aiming to revive the project.

Safeguarding the health of natural systems (Photo courtesy: Sally Tsoutas, WSU)

PART 2:
THE BIONARRATIVE IN A NUTSHELL

I have put the view that the survival of civilisation and future wellbeing of humankind will require a wave of new understanding spreading across the cultures of the world – understanding the story of life and the human place in nature. This story is of overarching significance for every one of us and for society as a whole.[8] Yet it is known and understood by only a minority of the human population. If this worldview were embedded at the core of the dominant cultures worldwide, the future prospects for humanity and the biosphere would be much brighter.

Why is this story so important?

First, the story of life conveys a sense of perspective crucial for understanding the human situation on Earth today. It reminds us that we are living beings, products of several billion years of biological evolution, and totally dependent on the processes of life, within us and around us, for our wellbeing and survival. Keeping these processes healthy must be our top priority, because everything else depends on them.

Second, this story makes it abundantly clear that the survival of civilisation will require big changes in the scale and kind of human activities on Earth.

Here is an extremely short version.

The first 4 billion years

Our planet is about 4600 million years old. The sun provides it with a constant supply of energy, in the form of visible light, and ultraviolet and infrared radiation.

The earliest living organisms on Earth are believed to have come into existence around 4000 million years ago. They were single-celled microbes, and they were the most complex form of life for over 1000 million years.

---

8   Note: For my own short version of the story of life, see S. Boyden. 2016. *The Bionarrative: the story of life and hope for the future.* ANU Press. Canberra. https://press.anu.edu.au/publications/Bionarrative.

TABLE 4: LIFE ON EARTH

Years ago	Event
4000 million	Earliest living organisms in existence, Bacteria and Archaea
2800 million	Microbes capable of photosynthesis
1500 million	First nucleated cells. Fungi probably in existence. Sexual reproduction?
600-700 million	First multicellular organisms – seaweeds, sponges, jellyfish, corals, worms, molluscs, sea urchins, starfish, trilobites
500 million	Earliest vertebrates – jawless fish (*Agnatha*)
450 million	Some plants and fungi move onto the land, soon to be followed by arthropods and lungfish
370 million	First amphibians
300 million	The heyday of the amphibians
200 million	Fewer amphibians. Their place taken by reptiles, including the earliest dinosaurs
200 – 65 million	The age of the dinosaurs, over 1500 species
65 million	A great crisis in reptilian history. Dinosaurs, flying reptiles and many other forms of life become extinct. Surviving vertebrates included some reptiles and birds, as well as a small group of tree-dwelling primates that looked something like present-day shrews. Among them were the ancestors of humankind
6 million	Some much larger primates walking in the African savannah with an upright posture, *Australopithecus afarensis*. A skull very like that of a chimpanzee, with a brain of around 500 cm^3
2.5 million	Primates in Africa making stone tools, *Homo habilis*. 90-120 cm tall. Brain about 800 cm^3. They consumed both plant and animal food
2 million	Primates upright walking, *Homo erectus*. 145-185 cm tall. Flat face like modern humans, and a prominent nose and brow ridges. Variable brain size, average about 1000 cm^3
300,000	*Homo sapiens* in Africa, brain size 1350-1450 cm^3. *Homo neanderthalensis* in Europe
100,000	Apart from the Neanderthals and *Homo sapiens*, at least three other kinds of humans are known to have been living outside Africa at about that time
12000	Some groups of *Homo sapiens* start farming
9000	First townships
250	Beginning of the Exponential Ecological Phase of human history

Fungi may have been in existence 1.5 billion years ago, and they were much in evidence 600 million years ago. Like animals, they get their energy and carbon from other organisms.

The first multicellular organisms came into being 600-700 million years ago. They included seaweeds, sponges, jellyfish, corals, worms, molluscs, sea urchins, starfish, lamp shells, and trilobites.

Since that time, biological evolution has resulted in the coming and going of a myriad of life forms, leading to the rich network of interacting and interdependent organisms that exist in our world today.

The earliest vertebrates were jawless fish (*Agnatha*) that lived around 500 million years ago.

The plants of the oceans have changed little since that time. In contrast, spectacular evolutionary changes took place among animals in the aquatic environment. By 200 million years ago the trilobites had entirely disappeared and were replaced by a new group of molluscs known as ammonites. At one time there were over twenty different families of ammonites, and some of them had a diameter of at least a metre. But the ammonites were also extinct by 60 million years ago.

Meanwhile there was remarkable diversification taking place among the bony fishes, leading eventually to the immense variety of fish species found in ponds, streams, rivers, lakes and oceans today.

By 450 million years ago, and possibly before this time, some plants and fungi were moving onto the land masses of the planet, soon to be followed by various kinds of arthropods. Vertebrates, in the form of lungfish, were venturing onto land 400 million years ago.

The early land plants evolved from a group of green algae. They included mosses, ferns and gymnosperms, like cycads, gingkos and conifers, followed much later by the angiosperms (flowering plants).

The heyday of the amphibians was around 300 million years ago. By 200 million years ago, their numbers had declined dramatically, and their place had been taken by reptiles. Birds and mammals evolved directly from the reptiles, which in turn had evolved from the amphibians.

The reptiles included the dinosaurs, which thrived from 200 to 65 million years ago. There were many different kinds of dinosaur, adapted to different kinds of habitat. Several aquatic groups evolved, some of which looked very much like fish, although they did not have gills, and they breathed air through a respiratory tract. There were also various forms of flying reptiles, with wings that consisted of leathery membranes, supported and extended by very elongated fingers. Some of them had a wingspan of over 7 metres.

The earliest mammals came into existence about 200 million years ago, at about the same time as the dinosaurs were emerging, and there were animals very like modern echidnas living around 150 million years ago. Mammals remained a rather insignificant group during this period of reptile dominance.

The first true flowering plants emerged about 160 million years ago, and since that time they have undergone spectacular diversification. They are now the dominant division of plants and are made up of two main groups, the monocotyledons and dicotyledons. In the monocotyledons, which include grasses, lilies, irises and crocuses, the seedling has a single leaf and the stems do not thicken. The seedlings of dicotyledons have two leaves and the stems become thicker as the plant matures.

Around 65 million years ago a great crisis occurred in reptilian history and many forms became extinct, including all the dinosaurs and flying reptiles and most of the large marine reptiles. This wave of extinctions is thought to have been caused by the impact of a massive comet, or asteroid, landing on the Earth. Many other forms of life disappeared at this time, including various microscopic foraminifera in the oceans and many aquatic animals, including the ammonites. Placental mammals, birds, lizards, snakes, turtles, crocodiles, fishes and plants were relatively unaffected.

Estimates of the number of different species alive today are very variable. The most widely cited estimate is 8.7 million, although some authorities believe the number is much greater than this.

Some common features shared by many animal species go back a very long way in evolution. The mouth and anus were in existence 600 million years ago. Among vertebrates, two eyes, two ears, a heart and a stomach go back at least 550 million years; and four limbs, each with 4 or 5 digits, go back to the earliest amphibians.

Photosynthesis

Around 2800 million years ago, micro-organisms emerged that were capable of photosynthesis – the process by which chlorophyll converts light energy from the sun into food energy. All animal and plant life on Earth today, and of course, human civilisation, are completely dependent on photosynthesis in green plants. Photosynthesis involves the uptake of carbon dioxide and water from the environment. It also results in the release of free oxygen into the atmosphere, making it possible for life forms to evolve that relied on oxygen for their respiratory processes. Some of this oxygen becomes converted into ozone (O_3), which floats up to the stratosphere where it acts as a filter, protecting the Earth's surface from ultraviolet radiation from the sun. As a result, by the time that humans appeared on Earth, and probably by two thousand million years before this, only about half of the total solar ultraviolet radiation, and a much smaller fraction of the short-wave UV-B rays, penetrated through to the Earth's surface. Had it not been for this effect, life as it exists on land today would not have been possible.

Although excessive UV radiation is damaging to living organisms, the ultraviolet rays that do still penetrate through the ozone layer play a number of useful biological roles, including the promotion of the synthesis of Vitamin D in human skin.

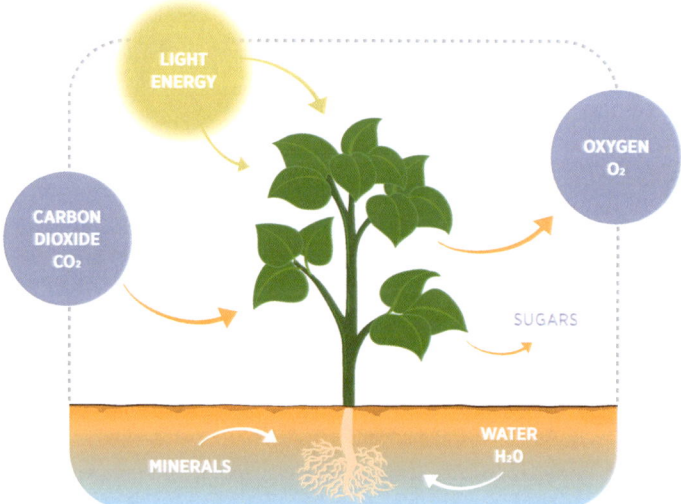

Figure 3. Process of photosynthesis

Reproduction

Universal among cellular organisms is the means by which genetic information is passed from parents to their progeny, providing the instructions that result in the new organisms developing and functioning as members of the species to which their parents belong.

The essential agent in this process is the genetic material of the cell, deoxyribonucleic acid (DNA). In animal and plant cells, chains of DNA are located in the cell nucleus, and in this situation DNA itself is capable of self-replication. It contains, in coded form, most of the information necessary for the formation of the new individual.

The inheritable characteristics of all organisms are determined by the arrangement of four nucleotides (cytosine, thymine, adenine, and guanine) in the genes, which are discrete areas or regions on the DNA chains.

Almost universal among plants and animals is the involvement of the sexual process at some stage in the reproductive cycle. This consists of the fusion of two separate cells (gametes) which, in the case of multicellular organisms, usually come from two different individuals, although in some species, coming from different parts of the same individual. In some very simple organisms, the two gametes may be identical, but in all higher species of plants and animals they are clearly different. The male gamete, or sperm, is mobile. The female gamete, or ovum, is larger and it is not mobile. During the formation of the gametes, through a process known as meiosis, the gametes lose half their genetic material. The fusion of the two cells results in the new fertilised egg, or zygote, which contains twice the amount of DNA contained in each of the gametes, half of it coming from the ovum, and half from the sperm.

The fertilised egg thus contains genetic material from two different individuals. Since it is very unlikely that the material from each parent will be identical, it follows that the offspring will be genetically different, even if only slightly, from either parent.

The sexual process means that the genetic material in a population is being constantly reshuffled. From the evolutionary point of view, the importance of sexual reproduction lies in the fact that, unlike in asexual reproduction, the precise genetic make-up of the new individual is different from that of either parent. This has the effect of maximising the number of genetic combinations in the population, and so increasing the potential of the population to adapt to environmental change through natural selection.

While the mechanism of sexual reproduction explains the continual rearranging of genetic material in populations, it does not tell us how entirely new genetic characteristics come into existence. This happens through the process of mutation, which consists of a chemical change in a gene that is perpetuated when the gene replicates in cell division. The change then affects the particular characteristic of the organism for which the gene is responsible. Mutations are normally rare events, but their frequency can be increased by certain physical and chemical agents, such as ultraviolet light, radioactive radiation and mustard gas.

The great majority of mutations are deleterious, so that cells that carry them do not survive. Occasionally, however, a mutation arises which, by chance, increases the likelihood of the organism surviving and successfully reproducing in the habitat in which it lives.

The Earth's recycling system

An essential feature of life on Earth is the cycling of nutrients, including carbon, sulphur, nitrogen, water, phosphorus, and oxygen among others, that are taken up from the environment, built into the tissues of living organisms, and then eventually released again, through death and decay, to become available for incorporation into new life. These nutrient cycles are essential for the sustainability of life.

The fertility of soil is largely dependent on its organic content, which plays a vital role in the nutrient cycles. It consists of decomposing plant and animal matter and the microorganisms involved in this decomposition. It also contains numerous other micro-organisms, such as those participating in the fixation of nitrogen from the atmosphere. Some components of the decaying organic matter, like waxes, lignins, and fats are relatively resistant to decomposition, and together they form a colloidal substance known as humus. Humus has an important influence on the capacity of the soil to support plant life. Soil is thus a rich depository for carbon.

The animals in soil include nematodes, millipedes, mites, insects, earthworms, burrowing amphibians, reptiles and mammals. There are believed to be around a million species of nematodes, most of them living in soil. On land, life above ground and life below ground are inextricably linked. Photosynthesis takes place above ground, but the vast majority of photosynthesising plants have their roots in the soil, from which they derive essential inorganic nutrients and water. So, life above ground is entirely dependent on life below ground, and vice versa.

Extinctions

The history of life has been marked by a number of mass extinctions. They were caused by extreme temperature changes and catastrophic events like massive volcano eruptions and asteroids hitting the Earth.

The most severe mass extinction occurred 250 million years ago when around 95 per cent of all marine species and 70 per cent of land species were wiped out. A mass extinction occurred about 65 million years ago. Many forms of life disappeared, including all the dinosaurs and flying reptiles. Some groups of reptiles survived, including birds, snakes, lizards, crocodiles and turtles. Some mammals also survived. It is believed that more than 99 per cent of all species that have existed on Earth are now extinct.

Another mass extinction is at present underway, due to the activities of *Homo sapiens*.

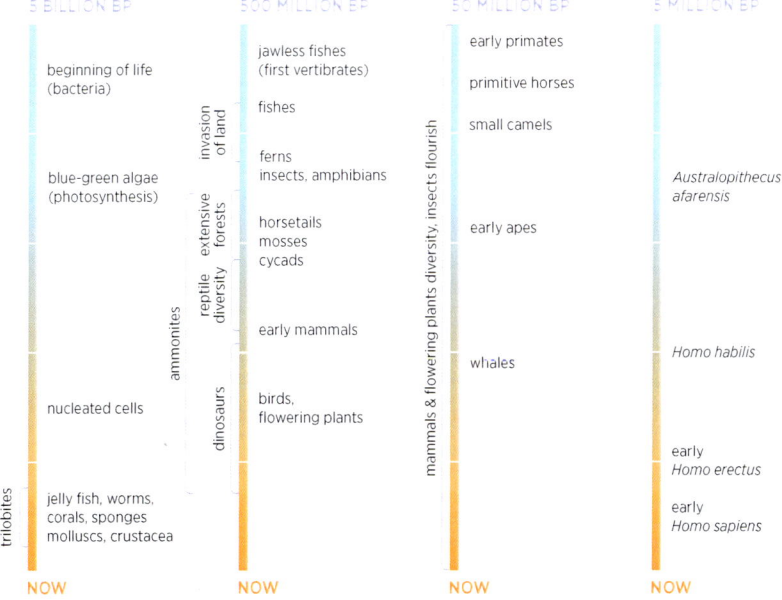

Figure 4: Some major developments in the history of life

Homo sapiens appears on the scene

During the last part of the dinosaur era, around 65 million years ago, there existed a small group of tree-dwelling primates that looked something like present-day shrews. Among them were the ancestors of humankind.

Five or six million years ago there were some much larger primates walking in the African savannah with an upright posture. One particularly well-preserved fossil is that of a young female found in Ethiopia, and dated about 3 million years ago. She is known informally as Lucy, and the species she belonged to has been called *Australopithecus afarensis*. She had a skull very like that of a chimpanzee, with a brain of around 500 cm^3.

Two and a half million years ago, and perhaps considerably earlier, there were primates in Africa making stone tools, marking the beginning of the so-called Paleolithic era of human prehistory, which lasted until the Neolithic era, which began around 12,000 years ago. One species, called *Homo habilis*, was 90-120 cm tall, and it had a brain with a volume of about 800 cm^3, which is about 300 cm^3 larger than the brain of a chimpanzee. These animals consumed both plant and animal food. Another rather similar species called *Homo rudolfensis* existed at about the same time.

After that, several different human species came into being, including *Homo erectus*.

The earliest 'modern humans', *Homo sapiens*, were living in parts of Africa around 300,000 years ago. They were tall people, with rounded skulls and steep foreheads, and their average cranial capacity was about 1,400 cm^3.

From about 200,000 years ago, and during most of the first part of the fourth, or Würm, glaciation, western Europe was occupied by a distinctive form of humanity classified as *Homo neanderthalensis*. The brains of adults ranged from 1,450cm^3 to 1,650 cm^3 in volume. They were well acquainted with the use of fire, they hunted big game and they dressed in animal skins. They used paints to decorate their bodies and sometimes they buried their dead.

Homo sapiens moved into Europe 40,000-50,000 years ago.

Apart from the Neanderthals and *Homo sapiens*, at least three other kinds of humans were living outside Africa at about that time. *Homo denisova* lived across Asia and *Homo longi* (Dragon man) was in China.

Although they are classified as a distinct species, both the Neanderthals and the Denisovans interbred with *Homo sapiens*. Dwarf hominids, called *Homo floresiensis*, existed on the island of Flores in Indonesia until about 50,000 years ago. They looked rather like *Homo habilis*.

Homo sapiens possesses an attribute now unique in the animal kingdom – the ability to invent, memorise and communicate with a symbolic spoken language. This aptitude for language led to the accumulation of shared worldviews, knowledge, beliefs, attitudes and technological knowhow in human groups. That is, it led to human culture.

Human culture has recently become a powerful force in nature. Shared worldviews, assumptions, and priorities, through their influence on human behaviour, have major impacts on other living organisms and on ecosystems. There is constant interplay between human culture and other aspects of biological systems.

Culture has led to activities that have been to the benefit of humans: these can be referred to as cultural adaptations. It can also lead to activities that have been greatly to their disadvantage, referred to as cultural maladaptations (see page 41).

The history of *Homo sapiens* has consisted of four quite distinct ecological phases. These phases can coexist in different parts of the world. For example, even today some of the Hadza people in northern Tanzania live as hunter-gatherers.

Paleolithic rock art in northern Spain (Photo: Museo de Altamira and D. Rodríguez)

TABLE 5: *HOMO SAPIENS* ON EARTH

Years ago	Event	Human Population
300,000	Earliest *Homo sapiens* in Africa	
180,000	Some *H. sapiens* move out of Africa	
60,000	*H. sapiens* arrives in Australia	
45,000	*H. sapiens* moves into Europe, displacing, or interbreeding with, Neanderthals	
12,000	Humans start farming in the Fertile Crescent	5 million
9000	Early townships (e.g. Çatal Huyuk)	
7000-5000	Some big cities in Mesopotamia	
6,800	Rice cultivated in China	
5000	Some big cities in Peru	
220	Beginning of industrial revolution	1 billion
84	World War 2 begins	
78	Nuclear weapons dropped on Hiroshima and Nagasaki, World War 2 ends	
50	Beginning of the computer age	
40	Beginning of the Internet	
Now	Anthropogenic climate change Thousands of weapons of mass destruction in existence Significant loss of biodiversity	8 billion

Phase 1 - The hunter-gatherer phase

This was by far the longest of the four ecological phases, lasting from at least 300,000 years ago, until the present day.

As in the case of all other animal species living in their natural habitats, for most of the time most members of hunter-gatherer bands are likely to have been in a state of good health. Indeed, they had to be, in order to survive and successfully reproduce under the demanding conditions of life. Because of the relatively low population density, people would seldom have suffered from such respiratory and enteric virus infections as common colds, COVID-19, influenza, gastric flu, measles, smallpox and German measles. Nor are they likely to have experienced bacterial infections like cholera and plague. However, infection with bacteria following physical injury would have been a constant hazard.

Ecologically, the most important culturally inspired activities in this phase were the deliberate use of fire and the manufacture and use of tools and weapons made out of wood and stone.

Phase 2 - The Early Farming Phase

Farming began in several parts of the world around 10,000 to 12,000 years ago. The global population is thought to have been about 5 million at that time.

Farmers, unlike hunter-gatherers, were no longer constantly on the move. Their lifestyles often involved long periods of hard physical work.

The introduction of farming marked a turning point in cultural evolution. It was a precondition for all the spectacular developments in human history that have occurred since that time.

Phase 3 - The Early Urban Phase

This phase began around 9,000 years ago, when fairly large clusters of people, sometimes consisting of several thousand individuals, began to aggregate in townships. Many of these people played little or no part in the gathering or production of food. Occupational specialisation became the hallmark of urban societies.

Although the new conditions offered protection from some of the hazards of the hunter-gatherer lifestyle, infectious disease and malnutrition became more important as causes of ill health and death.

Phase 4 – The Exponential Phase (the Anthropocene)

Ecological Phase 4 has recently come to be referred to as the Anthropocene, and because of the popularity of this term, I will use it in the rest of this booklet.

This phase was ushered in by the industrial revolution, which began about 250 years ago. It has been associated with profound changes in the ecological relationships between human populations and the rest of the biosphere.

The Anthropocene has seen an astounding profusion of technological innovations – from steam engines and motor vehicles to intercontinental rockets and spacecraft – and from electric lights, telephones and radios to thermonuclear bombs, computers, smartphones and the Internet.

This ecological phase has been associated with a massive and continuing growth in the human population, and an even more explosive increase in resource use and waste production by humankind.

Cultural maladaptations in the Anthropocene are on a scale and of a kind that threaten the whole of humankind, as well as countless other species.

Population and energy

There are now over 8 billion people on Earth, which is about 1,600 times as many as there were when farming began. Nearly 90 per cent of this increase has occurred in Ecological Phase 4 (**Figure 5**). The global population is now increasing at the rate of 1.4 million per week. I share the view of many ecologists that the number of people on Earth today greatly exceeds the maximum number that would ensure the long-term ecological sustainability of society and a reasonable quality of life for humankind.

There has also been a massive intensification of use of resources and energy and discharge of wastes. Humankind is now using around 20,000 times as much extrasomatic energy as was the case when farming began, and it is responsible for the emission of about 10,000 times as much of the greenhouse gas carbon dioxide (**Figures 6 and 7**).[9] More than 90 per cent of these increases have occurred since 1900. Nearly 80 per cent of the extrasomatic energy used today comes from fossil fuels.

---

9   Extrasomatic energy is energy used outside the human body, as opposed to somatic energy, which is energy used within the human body.

THE HUMAN POPULATION OF THE LAST 12,000 YEARS

Figure 5: Human population growth
Source: HYDE (2017); Gapminder (2022); UN (2022)

WORLD EXTRASOMATIC ENERGY USE FROM 1800

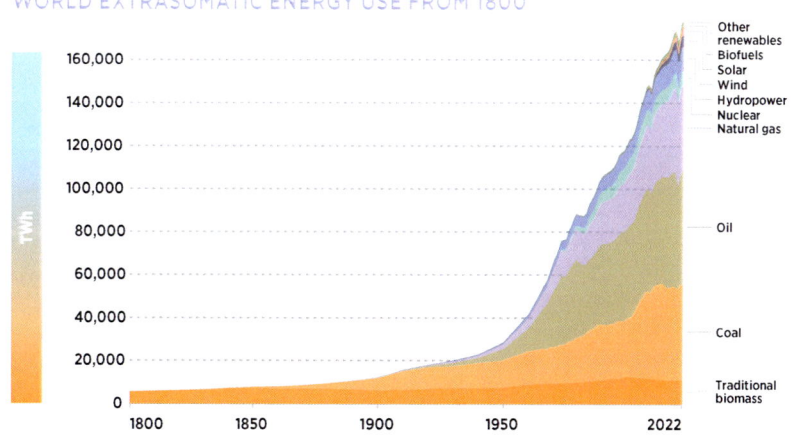

Figure 6: Human energy consumption
Source: Energy Institute Statistical Review of World Energy (2023); Vaclav Smil (2017)

Greenhouse gases

If it were not for certain gases occurring naturally in the atmosphere, the world's average temperature would be 33°C colder than it is. The average temperature would be around minus 19°C, instead of plus 14°C. This is because these gases trap some of the infrared radiation that escapes from the Earth's surface. This blanketing effect results in the lower layers of the atmosphere being warmer, and the upper layers colder, than if these gases were not there. This phenomenon is known as the natural greenhouse effect.

Water vapour is responsible for about 80 per cent of the natural greenhouse effect. The remainder is due to carbon dioxide, methane, and a few other minor gases. Carbon dioxide (CO_2) is responsible for about 15 per cent of the natural greenhouse effect. Were it not for the CO_2 in the atmosphere, the Earth's average temperature would be 5°C cooler than it is.

For the first 99.9 per cent of the history of *Homo sapiens*, the mixture of these greenhouse gases in the atmosphere was relatively constant. During the past two hundred years there has been an increase in the CO_2 concentration – from around 280 parts per million (ppm) in 1800 to 420 parts per million in 2023. The increase in atmospheric CO_2 is mainly the result of two sets of human activities, namely deforestation and the combustion of fossil fuels.

The amount of carbon dioxide emitted by the human population today is around 10,000 times greater than it was when farming began some 450 generations ago, and 90 per cent of this increase has occurred over the past 100 years (**Figure 7**).

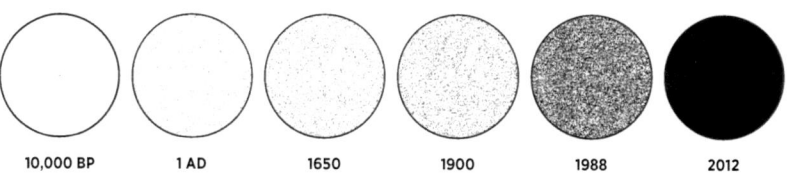

10,000 BP　　1 AD　　1650　　1900　　1988　　2012
One dot = 3.5 X 10^6 tonnes of carbon dioxide per year

Figure 7: Carbon dioxide production by the human species

The increase in atmospheric carbon dioxide is said to be responsible for 53 per cent of the current global warming. Humankind has also caused the release of several other greenhouse gases, most notably methane, halogenated compounds, tropospheric ozone and nitrous oxide.

Methane (CH_4) is generated by activities like farming livestock, especially ruminants, sewage treatment, distribution of natural gas and oil, coal mining and fossil fuel use. It is responsible for 15 per cent of current global warming. Halogenated compounds, like CFCs, HFCs and PFCs, are used in various industrial processes. They contribute 11 per cent of current warming. Tropospheric ozone (O_3) is given off during the combustion of fossil fuels, and it also contributes 11 per cent. Nitrous oxide (N_2O) comes mainly from use of fertilisers and use of fossil fuels, and also contributes about 11 per cent of current warming.

As a consequence of this anthropogenic increase in greenhouse gases in the atmosphere, by 2022 the Earth's average surface temperature had increased by about 1.2°C since 1880. This is referred to as the enhanced greenhouse effect. Sea levels are rising, and there is an increasing frequency of extreme weather events worldwide, such as powerful storms, typhoons, floods, droughts and heatwaves.

If governments do not take strong action in the immediate future, the consequences for humanity will be very serious indeed.

Deforestation

Deforestation of tropical forests is occurring at an ever-increasing rate – mainly to make way for pastures for beef cattle and oil palm plantations. Only about 6 million km² remain of the original 16 million km² of tropical rainforest that formerly existed on Earth. Over 30 million acres of forests are lost every year due to deforestation.

Deforestation is an important influence in global warming. It has been estimated that more than 1.5 billion tonnes of carbon dioxide are released to the atmosphere every year due to deforestation.

Farming

The United Nations Food and Agricultural Organisation warns that the world's agricultural systems face the risk of progressive breakdown of their productive capacity, due to excessive population pressure and unsatisfactory farming practices.

About 25 per cent of farmland worldwide has been degraded. If this trend continues, 95 per cent could be degraded by 2050. Land degradation is an important contributor to climate change, due to release of carbon and nitrous oxides.

As it is, 8 million people across the globe are underfed.

Farming is a leading cause of environmental destruction. It spreads over an area thirty times greater than that covered by urban sprawl, and it is the main cause of the current very high rate of extinction of animal and plant species.

Farming is also a cause of pollution of waterways with fertilisers, pesticides and other chemicals.

Techniques are now available for producing meat by culturing muscle and fat cells from animals in the laboratory. Meat produced in this way is called cultured meat. Small quantities of cultured beef and chicken are now available commercially in several countries.

Microorganisms are another possible food source for the future. In Finland, a company called Solar Foods uses renewable electricity to split water into oxygen and hydrogen. The hydrogen is then used, along with carbon dioxide, minerals and mineral nutrients, as food for bacteria which multiply to form an edible protein-rich powder called Solein. It will be available commercially some time in 2023.

A strong body of opinion holds that protein-rich foods produced in the laboratory should replace farm-grown meat in the human diet, much to the benefit of the natural environment.

With regard to plant food, there is plenty of scope for increasing yields of green vegetables, roots and fruits in the urban setting, thereby lessening the impact on natural ecosystems.

Waste production and pollution

Plastics have been introduced for manufacturing a very wide range of objects. About 9 million tonnes of plastic waste are discharged into the sea every year, and the amount has been predicted to double in 11 years. Environmental pollution with discarded plastics is causing a dramatic decline in populations of many seabirds. 5000 – 15,000 turtles become entangled in discarded fishing gear every year. According to one prediction, by the year 2050 there will be more plastic in the oceans than fish.

St. Rollox Chemical Works at the opening of the Garnkirk and Glasgow railway (Artwork: David Octavius Hill, c. 1831)

The release into the atmosphere of chlorofluorocarbons (CFCs) and halons, which are gases formerly found in aerosol spray cans and refrigerants, has resulted in some thinning of the ozone layer in the stratosphere – causing an increase in UV radiation reaching the Earth's surface. In 1987 an international agreement, the Montreal Protocol, was signed, designed to protect the ozone layer by phasing out the production of ozone-depleting substances. As a result of this international agreement, the ozone hole is slowly recovering. It is believed that it will return to 1980 levels between 2050 and 2070.

In 1962 Rachel Carson drew attention to the widespread and destructive ecological impact of DDT and other pesticides. DDT belongs to a family of synthetic compounds known as Halogenated Persistent Organic Pollutants (POPs) that are used as pesticides as well as in certain technological processes. POPs have been found to accumulate in the internal organs of living creatures and are believed to be responsible for increasing and widespread infertility in wild animals, and possibly also in humans. They may also contribute to an increase in breast cancer and reduced sperm counts in men. POPs are very persistent in the natural environment, and they have been found in the organs of animals in areas as far away from where they were released as the Arctic and Antarctic.

Three to four million tonnes of heavy metals, solvents, toxic sludge and other waste is dumped into the world's rivers and oceans every year. Urban air pollution, due mainly to the combustion of fossil fuels, is a significant cause of ill health and death in many cities worldwide, especially in Asia. It is estimated that 2.5 billion people are exposed to air pollution levels seven times WHO guidelines.

Climate change and the oceans

Increasing concentrations of greenhouse gases in the atmosphere and consequent climate change are having major impacts on the oceans. These impacts include increase in temperature, acidification, rise in sea levels and changes in ocean current patterns. All these factors have complex and unpredictable effects on marine ecosystems.

Spacecraft

In 1957 the USSR launched the first artificial satellite into space. It was named Sputnik 1, and it was in orbit for three months. There are now, at the time of writing, over 4,800 active satellites in orbit, and several thousand dead satellites, as well as many thousands of pieces of space junk large enough to be tracked.

Loss of biodiversity

The current rate of loss in biodiversity resulting from human activities is estimated to be 100 to 1000 times higher than the natural background extinction rate. According to some estimates, one million of the world's species are now under threat of extinction, and 25 per cent of all mammal species could be extinct in 20 years.

Environmentalism and the Green Movement

During the past half century there has been increasing awareness in some sections of the community that our present society is causing serious damage to the ecosystems of our planet. Environmentalism has emerged as a new ideology. The advent of the Greens as a political entity is an indication of this growing concern, although election results suggest that it is still shared by only a minority of the population.

There are also countless groups focusing on specific ecological issues. To mention but a few local examples here in Australia, we have the Climate Institute, Sustainable Population Australia, SEE Change groups, Greenpeace Australia, The Wilderness Society, Permaculture groups, Healthy Soils Australia, 350 Australia and Landcare groups.

At the international level there has been a series of major conferences on sustainability, organised by the United Nations, starting with the Conference on the Human Environment held in Stockholm in 1972. There have also been many important international conferences on specific ecological issues, including the United Nations conferences that led to the Paris Agreement on Climate Change in December 2015.

With the exception of the Paris Agreement, these warnings have not penetrated to the core of the prevailing cultures of the world. We have only to listen to the pre-election speeches of our political leaders for proof of this statement. Although some important measures have been taken here and there to protect aspects of the natural environment, they have not been allowed to interfere with the inexorable thrusts of ever-moreism and market forces. The juggernaut rolls on.

An interesting and important development in the public health arena has been the Planetary Health Movement. A commentary in the medical journal The Lancet in March 2014 called for the creation of a movement to broaden the field of public health, so that it includes the health of the natural ecosystems on which we depend. In 2015, the Rockefeller Foundation and The Lancet launched this concept as the Rockefeller Foundation-Lancet Commission on Planetary Health. In December that year, Harvard University, together with some other organisations, founded the Planetary Health Alliance to promote the concept. Funded by Rockefeller Foundation, the Alliance plans to support the development of a "rigorous, policy-focused, transdisciplinary field of applied research of the structure and function of Earth's natural systems".

So, while the process of cultural reform is certainly underway, it has a long way to go; and the inevitable counter-reform backlash is very much in evidence. The ecologically maladaptive assumptions of the prevailing culture remain firmly entrenched, and the reform process is in need of a big boost.

Conclusion

The story of life on Earth reminds us that we humans are both a product and a part of nature, and that we are totally dependent on the processes of life, for our survival and wellbeing. It also leads us to the inescapable conclusion that the present pattern of human activities on Earth is not sustainable ecologically. The days of Ecological Phase 4 are numbered.

PART 3:
SOME BIOHISTORICAL PERSPECTIVES

HUMAN CULTURE: A NEW KIND OF FORCE IN THE BIOSPHERE

The human capacity for culture was certainly of major biological advantage in the evolutionary environment of our species, and in more recent times it has resulted in countless benefits for our species.

Apart from its practical advantages, culture adds richness to human experience. It did so in the lives of hunter-gatherers – as in storytelling, musical traditions, dancing and other forms of artistic expression. It does so today in so many ways. Culture makes a huge contribution to the sheer enjoyment of life.

But there is another side to the picture. The consequences of our capacity for culture are not all good. In fact, as cultures evolve, they often come to embrace not only factual information of practical value, but also ideas and assumptions that are sheer nonsense, leading to behaviours which are equally nonsensical.

Sometimes these cultural delusions have resulted in activities that have caused a great deal of unnecessary human distress, or damage to ecosystems, or both. We refer to culturally inspired activities with these characteristics as cultural maladaptations.

There are countless instances of cultural maladaptation in the early farming and early urban phases of human history. A particularly tragic example was the ancient Chinese custom of foot-binding, which prevented the normal growth of the feet of young girls and caused them excruciating pain. This particular cultural maladaptation was mutely accepted by the mass of the Chinese population for some forty or more generations. Such is the brainwashing power of culture.

Only a few generations ago, imperialism and the slave trade were completely acceptable in the eyes of the prevailing cultures of the Western world.

Cultural developments can be adaptive in some respects and maladaptive in others. The use of fossil fuels to power motor vehicles could initially be seen as a cultural adaptation, although in the longer term it is seriously maladaptive.

Fortunately, humans have the ability, through their capacity for culture itself, to bring culture back on track when it goes off the rails.

Nowadays, when some people declare the biological or social consequences of culturally inspired activities as undesirable, a period of discussion and debate ensues about the causes of the problem and possible remedies. Eventually new understanding can bring about modifications in cultural assumptions and priorities, leading to appropriate changes in human activities. This process is referred to as cultural reform.

Cultural reform is often quite complicated, involving prolonged interactions between different interest groups in society. A key role is often played initially by minority groups, occasionally by single individuals, who start the ball rolling by drawing attention to an unsatisfactory situation. An example is Rachel Carson who, in her ground-breaking book Silent Spring, drew attention to the insidious and destructive ecological impacts of certain synthetic pesticides.

Almost invariably these expressions of concern coming from reformers are promptly contradicted by others, the counter-reformers, who set out to block the reform process. This predictable backlash often involves, but is not restricted to, representatives of vested interests who believe that the proposed reforms will be to their financial disadvantage. They are likely to argue that the problem does not exist, or that it has been grossly exaggerated, and they try to ridicule the reformers by calling them alarmists, fanatics, scaremongers or prophets of doom. Nowadays some of the counter-reform forces are extraordinarily powerful.[10]

Eventually, if the reformers are successful, a change comes about in the dominant culture and members of governmental bureaucracies and other organisations set about working out ways and means of achieving the necessary changes. Their efforts may still be hindered by the stalling tactics of counter-reformers.

The so-called Enlightenment

Towards the end of the 17th century and during the 18th century the intellectual movement commonly, but misguidedly, referred to as the Enlightenment, was underway in Europe. This movement emphasised rational thought, as opposed to religious tradition, as a means of understanding the universe and making things better for humankind.

---

10 For a detailed discussion in the context of tobacco smoking, CFCs and climate change – see *Merchants of doubt* by N. Oreskes and E.M. Conway (2010).

I say 'misguidedly' because a more appropriate term would be Partial Enlightenment. Its great weakness lay in its association with the idea that nature is out there to be conquered.

Francis Bacon is credited with originating the idea of improving the human condition by conquering nature, and Descartes believed that we should become 'like masters and possessors of Nature'.

Does it make sense to set out to conquer the living system that gave rise to us, of which we are a part, and on which we are totally dependent? No, it does not; but it does make sense to try to understand it, to respect it, and to seek to live in harmony with it.

The tyranny of culture and cultural gullibility

The Bionarrative alerts us to the brainwashing power of culture. It warns us to be constantly vigilant – making sure that the worldviews, assumptions and priorities of our cultures are in tune with reality; and that they are not leading us to behave in ways that cause unnecessary distress to humans or other animals, or that cause damage to the living systems of the natural environment.

The tendency of humans to blindly accept the assumptions and prejudices of the cultural soups in which we have been immersed since childhood lies behind most of the conflicts between different religious and ethnic groups in our world today. Cultural gullibility is a fundamental, and potentially very dangerous, human characteristic.

HEALTH AND DISEASE

Evolution and human health

The health needs of animals, including *Homo sapiens*, are determined by their evolutionary background. This is because, through the processes of evolution, species have become well adapted in their innate biological characteristics to the conditions prevailing in the environment in which they are evolving.

If an organism is exposed to conditions of life that differ significantly from those to which it has become adapted in its natural environment, it is likely to be less well adapted to the new and different environment, and it is likely to show signs of maladjustment. It will be less healthy than in its natural environment. This fundamental evolutionary health principle applies to all plants and animals.

The evolutionary health principle clearly applies to a wide range of physical aspects of life conditions in humans. There is no diet better for humankind than that which was typical for hunter-gatherers. It is also clear that the principle is applicable to some aspects of behaviour. Marked deviations from natural sleeping patterns cause maladjustment, and health is likely to be impaired if levels of physical exercise deviate markedly from those in the natural habitat.

There are good reasons for supposing that the evolutionary health principle also applies to psychosocial and relatively intangible aspects of life experience. For example, the conditions of life of hunter-gatherers are usually characterised by a sense of purpose in daily activities and plenty of convivial social interaction. Most of us would agree that such conditions are likely to promote health and wellbeing in our own society.

Taking our knowledge of the conditions of life of hunter-gatherers as a starting point, we can put together a working list of physical and psychosocial conditions likely to promote health and wellbeing in our species (see page 12). They are referred to as universal health needs, because they apply to all members of the human species, wherever or whenever we may be living.

Not every item on the psychosocial list is absolutely essential for health. Lack of satisfaction of one psychosocial health need may be offset by the satisfaction of others. On the other hand, every item on the list will, if satisfied, make a positive contribution to health and wellbeing.

Stressors and meliors

Stressors are experiences which, if prolonged, cause anxiety and distress, and they are a normal aspect of life. If they are short-lived and not too severe, they can be seen as contributing positively to the quality of life and wellbeing; but if they are excessive and if they persist, they can interfere seriously with both mental and physical health. Equally important are experiences which have the opposite effect to stressors, and which give rise to a sense of enjoyment. Such experiences have been called meliors.[11] Common meliors include the experience of creativity, fun, aesthetic enjoyment, and conviviality.

---

11 Boyden, S., S. Millar, K. Newcombe, B. O'Neill, 1981. *The ecology of a city and its people: the case of Hong Kong.* ANU Press, Canberra. p 343

Every person can be considered at any given time to be at some point on a hypothetical continuum between a state of distress and a state of enjoyment. Their position on this continuum is largely a function of the balance between meliors and stressors in their recent experience.

The cultural environment has an immense influence on the levels and kinds of meliors and stressors in an individual's daily experience. Culture also influences the environmental costs of avoiding stressors or experiencing meliors.

Parasites and infectious disease

All animals and plants live in intimate association with other organisms of different species. In some cases, these associations are of benefit to both organisms as, for instance, in the case of lichens. Each lichen consists of an organised network of filaments of a fungus, and entangled in this network are cells of algae. The algae carry out photosynthesis and so contribute large energy-containing food molecules to the complex, while the fungus provides support and absorbs water and soluble nutrients from the environment.

The word symbiosis is used to describe mutually beneficial associations of this kind. Such associations may involve animals or plants, and they are sometimes obligatory, sometimes optional, sometimes permanent and sometimes transient. In the case of the lichens, the algae can grow independently, but the fungus cannot.

Parasitism is a type of association between two organisms in which one of them, the parasite, is dependent on, and lives at the expense of the other, the host. The host provides the parasite with a habitat and with nourishment. Internal parasites live within the host's body, as in the case of the parasitic worms that are found in the intestines of animals, and various bacteria that live and multiply in internal organs. External parasites, like the fleas of mammals and the mistletoes of plants, live on the outside surface of the host, but still derive nourishment from it. Parasitism is extremely common, and there would not be a single species of multicellular animal or plant that does not normally harbour parasites of one kind or another. The parasites of mammals include not only many kinds of single-celled organisms, but also roundworms, tapeworms, hookworms, liver flukes, mange mites, lice, ticks and fleas.

While all parasites feed on nutrients supplied by their hosts, the damage they cause is variable. Under typical natural conditions animals are usually not seriously disadvantaged by the parasites they carry; but unnatural crowding, or ill health from other causes, often leads to damaging levels of parasitic infestation.

The great majority of bacteria, protozoa and fungi are free-living and incapable of multiplying in the bodies of living animals and plants.

Infectious disease

Infection with parasitic micro-organisms often causes signs of overt disease. Well-known examples in plants include potato blight and wheat rust, both of which are due to a fungus. Examples in humans include malaria and dysentery, which are due to protozoa, and tuberculosis and cholera which are due to bacteria.

Most disease-producing bacteria, protozoa and fungi are incapable of multiplying outside the bodies of their hosts. However, there are a few micro-organisms which are normally free-living, but which can, under certain circumstances, multiply in animal tissues and cause disease. An example is the bacterium *Clostridium tetani*, which lives naturally in the soil. If this organism gains access to the body of an animal through a wound and becomes surrounded by dead tissue, it may be able to multiply. When it does so, it produces a protein that is extremely toxic for most mammals, causing the symptoms of tetanus.

Some of the more severe infectious diseases of plants and animals are caused by viruses, most of which are not visible under the light microscope. They range in size from the virus of foot and mouth disease which has a diameter of only 21 nanometres, to cowpox virus, which measures 210 x 260 nanometres (a nanometre is one billionth of a metre!). Most bacteria measure 1000 to 2000 nanometres. Viruses are relatively simple structures, with a central core of nucleic acid, which is usually surrounded by a layer of protein. They are only capable of multiplying within living cells.

The presence of a virus in the cells of a host does not necessarily cause any serious harm, and viruses can sometimes lie latent in the body tissues for long periods without giving rise to any symptoms. On the other hand, some viruses cause severe disease. In humans, infectious diseases due to viruses range from relatively mild conditions like the common cold and gastric flu to poliomyelitis and smallpox.

Infectious disease became much more common as a cause of ill health and death among humans when large numbers of people started living in close proximity in townships and cities.

Immunity

When the tissues of a mammal or a bird are invaded by microbes of a kind that the animal has not experienced previously there usually occurs an immediate inflammatory response, in which mobile cells, known as phagocytes, attempt to ingest and digest the intruding organisms. In cases when the microbes can withstand these mechanisms, a second phase of the defence process comes into play – the immune response. As a result, after about a week or ten days the animal's tissues become much more sensitive to this particular microorganism and its products. This increased sensitivity is associated with the appearance in the blood and other body fluids of antibodies, which are protein molecules that have the property of combining specifically with the macromolecules that form part of, or are produced by, the invading microorganism. Because of the presence of antibodies and some other changes in the host's tissues, the cellular reaction to the infectious agent is greatly enhanced and much more effective. In the case of cholera, for example, if infected humans live long enough for the immune response to develop, they are likely to survive and overcome the infection.

In most virus diseases the body's defence mechanisms bring the infection to an end quite quickly. In the more severe diseases, however, like poliomyelitis and smallpox, serious and lasting damage, and sometimes death, may come about before these mechanisms become effective.

The specific immunity produced in this way is sometimes long-lasting, as in the case of measles. In other diseases, like the common cold, it wanes after a year or so.

All artificial immunisation procedures are aimed at bringing about an immune response against disease-causing organisms, and so conferring protection against natural infection at a later date.

COVID-19

In December 2019 a new virus disease affecting the respiratory system was detected in the city of Wuhan in China. It later became known as COVID-19. It spread very rapidly across the world. By January 2023, COVID-19 had caused between 13-16 million deaths worldwide. In unvaccinated populations about 80 per cent of cases are mild, 14 per cent severe and 5 per cent critical. The death rate in unvaccinated populations is 0.5 to 1.0 per cent. Several vaccines became available in 2021 and 2022, and these have been very effective in reducing the number of severe cases and the mortality rate.

The risk of more pandemics in the future, possibly more severe than COVID-19, is very real. The likelihood of infectious diseases jumping from animals to humans is said to be higher than ever before.

RELIGION

An outcome of the capacity for culture from very early times was the emergence of religion as a universal feature of human society. Without exception, all recent hunter-gatherer groups believed in a supernatural, or spiritual, dimension of the universe, although there was enormous variation in the details of these belief systems. There is every reason to suppose that this has been the case for tens of thousands of years.

The early farmers all had their gods and spirits, although the details differed from one region to another. The dominant religious theme for several thousand years in south-western Asia and Europe was the notion of a Mother Goddess or 'Female Principle', who was worshipped as the giver of life. Rituals were practised aimed at pleasing the Goddess in the hope of improving the chances of good harvests, good health and successful reproduction.

In the early cities in Mesopotamia the religious sense of oneness with nature was abandoned. Each city state had its own god, who was now male, and conflicts between city states were viewed as being conflicts between the different gods.

Most of the religions of the Near East were polytheistic, and this was also true in Ancient Greece and Rome. However, Zoroastrianism, which was founded by the Persian prophet Zoroaster in the late 7th or 6th century BC, or possibly much earlier, was based on the idea of a continuous struggle between a single god of creation, goodness and light, and his arch enemy, the spirit of evil and darkness. Unlike some other early urban religions, it included a highly developed ethical code. Judaism was also based on belief in only one God.

The teachings of Buddha around 600 BC were initially relatively simple, as were those of Jesus of Nazareth. In neither case was this simplicity to last. In the case of Christianity, the processes of elaboration, intellectualisation and institutionalisation soon led to complicated sets of theories and rituals, with notable contributions from older religions and philosophies. A professional priesthood came into being, and ultimately the Christian Church split into a few large and sometimes mutually intolerant sects, and numerous smaller ones.

Another great religion of early civilisation, Hinduism, became extremely complex and involved the worship of a few major deities and countless minor ones. The various sects within Hinduism were relatively tolerant of each other, and of other religions. Similar elaboration occurred in the religions of the New World before the European conquest. In the case of the Toltecs, who dominated the valley of Mexico before it was overrun by the Aztecs, a relatively basic nature-worship became transformed into a very elaborate polytheism.

From the beginning, the great majority of humans have grown up and lived in cultural systems that clearly defined the nature of a supernatural world and spelled out their religious obligations. Although there must always have been a few sceptics, most people never questioned the validity of the belief systems of their local culture. There were exceptions, of course, as when occasional individuals became suddenly converted from one religion to another.

The sense of religious conviction was often extraordinarily strong, and throughout the history of civilisation differences in religious beliefs have been the cause of widespread human suffering. In Islam and Christianity, mutual intolerance has resulted in a great deal of bloodshed, as in the case of the Crusades and the 9/11 massacre in New York. Religious intolerance was not confined to the Western world. In China, religious persecutions around 845 AD are said to have resulted in the destruction of 44,600 Buddhist religious establishments and the enslavement of 150,000 Buddhist nuns and monks.

The assumption that in warfare one's own god is on one's own side has persisted for thousands of years. A Spanish eyewitness to the conquest of Middle America wrote in his diary: 'When the Christians were exhausted from war, God saw fit to send the Indians smallpox, and there was a great pestilence in the city'.[12]

Occasionally, of course, there have been individuals who have appreciated the insanity of religious bigotry. Akbar, a Mogul emperor of India from 1556 to 1605, was such a person. He became acutely aware of the absurdity of the whole multi-faceted and fragmented religious scene in India, and of the needless distress caused by religious intolerance. He refused to accept the idea that, because he was the conqueror and ruler and happened to

---

12 D. R. Hopkins. 1983. *Princes and peasants: smallpox in history*. University of Chicago Press, Chicago. p.201.

have been born a Mohammedan, that Mohammedanism was true for all humankind. It was his aim that all people, whatever their race or religion, should participate equally in India's public life.

Today the great majority of the world's population adheres to one religion or another. One recent estimate suggests that about 33 per cent of people are Christians, 19.6 per cent Muslims, 13.4 per cent Hindus, 5.5 per cent Buddhists. There are also countless different sects within the major religions, each with its own particular creed. Judaism, Sikhism, Zoroastrianism and Taoism are among numerous other smaller religious groupings, each followed by less than 1 per cent of the total human population. Only about 15 per cent of people were described as non-religious.

Religious intolerance is still very evident today, and it causes an immense amount of human distress. On the other hand, religion is also a source of great comfort for many people.

WARFARE

Deliberate intraspecific killing is not common among vertebrates. It occurs sometimes in chimpanzees and some other species, including wolves, big cats, and the banded mongoose, but large-scale lethal combat is a uniquely human characteristic.

Judging from evidence from recent hunter-gatherers, it seems likely that mortal conflict sometimes occurred between different human groups in the hunter-gatherer phase of human history, although it would not have been a constant feature of primaeval society.

Similarly, there have been plenty of farming communities that have lived at peace with their neighbours for long periods of time, and archaeological evidence suggests that the early farmers of the valleys of the Tigris and Euphrates Rivers, were not involved in warfare.

On the other hand, in some early farming societies violent hostilities with neighbours was an important feature of life. There is evidence that massacres of large numbers of people sometimes occurred in Europe in the early Neolithic period, and that around 5000 years ago farming people in the south of England built and attacked fortified settlements.

By around 5000 years ago, highly organised fighting between the city states in Mesopotamia, and between city states and barbarian raiders, was commonplace.

For centuries history books have extolled the prowess of men who commanded armies that succeeded in annihilating large numbers of perceived enemies.

An example is provided by the words of the Greek historian Plutarch, who writes in the following glowing terms of Julius Caesar:

> *Caesar surpassed all other commanders in the fact that he fought more battles than any of them and killed greater numbers of the enemy. For, though his campaigns in Gaul did not last for as much as ten complete years, in this time he took by storm more than 800 cities, subdued 300 nations and fought pitched battles at various times with three million men of whom he destroyed one million in the actual fighting, and took another million prisoners.*[13]

The professional soldier came to be accepted as a natural and necessary component of most urban societies. For millennia the sword held pride of place among human artefacts as the symbol of masculine virtue.

Warfare was not, however, an inevitable concomitant of urbanisation. The remains of the township of Caral in Peru, which came into existence around 5000 years ago, show no trace of warfare. No battlements and no weapons have been found. Similarly, excavations at ancient cities of Harappa and Mohenjo-daro in the valley of the Indus River in Pakistan have revealed no indication of military activity until the very end of their history.

One of the unfortunate consequences of the human capacity for culture has been the transmission of hatred across generations. It is a major determinant of aggressive behaviour in some parts of the world today.

Weaponry

Early in hominid history, our ancestors applied their tool-making prowess to the manufacture of weapons. In primaeval times these were used mainly for hunting animals for food, although they would sometimes have been used for fighting with other human groups. The weapons were broadly of two classes. First, there were close-range weapons, like clubs and hand axes. Second, there were projectile weapons, like stones and sticks, which were thrown at the target, initially by the human arm, but later by other means, as in the case of the bow and arrow. Spears were used as both close range and projectile weapons.

---

13 E. Canetti. 1973. *Crowds and power* (translated by C. Stewart) Penguin, Harmondsworth. p 269.

After the beginning of urban civilisation most weapons were designed especially for killing people, and they fell into the same two classes: close range and projectile weapons. When techniques of metallurgy were developed, spearheads, and sometimes spear shafts, were made of copper or bronze. The other important short-range weapon, invented and developed especially for cutting or thrusting into flesh, was the sword.

The discovery of the explosive potential of a mixture of saltpetre, sulphur and charcoal, otherwise known as gunpowder, is believed to have been made in China over a thousand years ago. In the mid-13th century Roger Bacon in England wrote a formula for gunpowder. The first cannons, which were made of bronze, were introduced at the beginning of the 14th century.

Cultural evolution in Europe and Asia has been associated with a progressive increase in the number of people actively participating in wars. In World War I about 53 million people were mobilised into the armed forces, and 8 to 10 million were killed.

In World War II aerial bombing of cities resulted in very large numbers of civilian casualties. The armed forces of the warring nations numbered about 30 million and the total number of individuals killed, military and civilian, was probably between 35 and 40 million.

Until recently, microbes have caused more deaths among warriors than combat itself. In the Crimean War (1854-1856), for example, around 60,000 soldiers were killed in action or died from wounds and 130,000 died from infectious diseases.

By the time of World War I some of the combat was no longer on a one-to-one basis. A single pull of a trigger of a machine gun could kill a dozen soldiers; and one artillery shell could destroy many individuals who were completely out of sight of the gunners.

This fundamental change in the nature of armed conflict had further progressed by the time of World War II, when various kinds of rockets were introduced. Technology now exists which makes it possible for one person to cause the death of millions of people thousands of miles away. At 8.13 am on 6th August 1945, a nuclear bomb was dropped from an American aircraft onto the city of Hiroshima in Japan. At least 140,000 people, about 40 per cent of the population of the city, were immediately killed, or died soon afterwards. Three days later, another nuclear bomb was dropped on the Japanese city of Nagasaki, and 26 per cent of its population of about 280,000 was killed outright. Bombs now exist with an explosive power a thousand times greater than that which was dropped on Hiroshima. Most commentators consider it likely that a major nuclear war today would leave some survivors, especially in the southern hemisphere. The effects of such a war on the planet's ecosystems are uncertain. It is possible that the biosphere as we know it would collapse, and would no longer be capable of supporting a human population.

So, for the first time in the history of life on Earth, and in the lifetime of many of us alive today, a single species of animal has developed the ability to destroy most, if not all, of its kind within a few days. It owes this achievement to its capacity for culture.

Mention must also be made of the enormous amount of effort and resources that have been devoted in modern societies to the development of other sophisticated weapons of mass destruction. Thus, apart from the advances in nuclear armaments, great progress has been made in the development and production of chemical and biological weapons.

In conclusion, there is clearly nothing in human nature that precludes lethal combat between different groups of people. On the other hand, there is also nothing in human nature that rules out the possibility of different human groups living permanently at peace with each other. A major determinant of whether or not warfare and terrorism continue to be features of civilisation will be the extent to which people allow themselves to be blinded by narrow, pernicious and maladaptive cultural delusions.

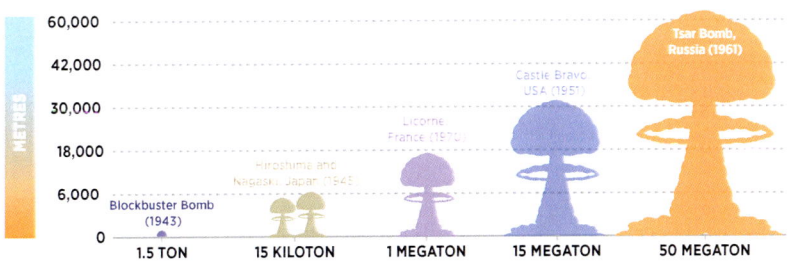

Figure 8: Increasing destructive power of bombs, 1943-1961

VEGETARIANISM & VEGANISM

Some people are vegetarians because they believe it is morally wrong to consume other animals. I am not one of them. Nature created our species as an omnivorous animal, just as it created deer as herbivores and lions as carnivores. For something like 12,000 generations my *Homo sapiens* ancestors have been meat eaters.

The cattle that graze on my family's farm owe their existence to the fact that humans eat meat. They have a good life, and they seem to be happy most of the time.

However, I feel strongly that we should completely change the slaughtering procedure. Animals should be killed instantly in the paddock or yard on the farm, and then transported to the butchery. None of this trucking of living animals long distances to sale yards, and none of the extreme fear as they are lined up for slaughter.

The ecological arguments for vegetarianism and veganism are more persuasive.

The farming of animals for food production is now on a scale that causes serious ecological problems worldwide. In some areas animal farming is leading to significant loss of biodiversity. This is especially so in the tropical rain forests of Central and South America, where deforestation to make way for cattle is resulting in the loss of many animal and plant species.

Land degradation caused by overgrazing is also a serious problem, especially in dry areas of the world. Impacts include biological impoverishment of the soil, soil erosion and eutrophication of streams and rivers.

According to the FAO, livestock, including poultry, account for 14.5 per cent of anthropogenic greenhouse gas emissions.

A paper in the journal, Science, has highlighted the scale and seriousness of the problem, leading the authors to advocate universal veganism to save our planet.[14] They estimate that a vegan world would produce 40 per cent less food-based greenhouse gas emissions, 50 per cent less acidification on land, 49 per cent less eutrophication.

---

14   J. Poore and T. Nemecek. 2018. Reducing food's environmental impacts through producers and consumers. *Science* 360. Issue 6392. Pp.987 – 992.

It would use 19 per cent less water; and it would cut land use by 76 per cent. They point out that there is big variation in the environmental impact of different farming practices. It has been reported that the world's 10 per cent worst beef producers emit 12 times more greenhouse gas, and take up 50 times more land, to produce a unit quantity of protein, compared to the best 10 per cent. Referring to this paper, George Monbiot of the Guardian, writes:

> *We can neither feed the world's growing population nor protect its living systems through animal farming. Meat and dairy are an extravagance we can no longer afford.*[15]

So, apparently, we have a sad situation. For some 300,000 years humankind fitted into the biosphere much like any other omnivorous species, and although there were fluctuations in population from time to time, there were never enough people to threaten the integrity of the ecosystems on which they depended. After the introduction of agriculture around 12,000 years ago, the human population began to increase and there are now 1,600 times as many people on Earth as there were when farming began.

The advocates of veganism argue that the population has now reached such a level that we must stop eating meat, which was such an important part of the natural diet of our species. We must shift to an unnatural diet to save the planet, simply because there are so many of us.

Not surprisingly, there has been a backlash from supporters of the meat industry. Critics suggest the paper in *Science* is too narrow. They ask: Why single out meat production, when there are so many other human activities threatening the integrity of the biosphere? Activities like producing palm oil and soya bean oil and the use of fossil fuels, and even keeping pet dogs and cats, make a huge contribution to the unsustainability of modern society. They draw attention to the fact that, in some parts of the world, populations depend on meat eating for their very survival, and that well managed grazing pastures can have a positive effect on biodiversity.

In my view, the information assembled in the Science paper is probably sound, and the production of meat and dairy products is making a major contribution to climate change and loss of biodiversity. This is, of course, a consequence of the massive increase in the number of people on Earth.

---

15   G. Monbiot. 2018. The best way to save the planet? Drop meat and dairy. Farming livestock for food threatens all life on Earth. *The Guardian Weekly.* 199. No.2 p.48.

BIOMETABOLISM & TECHNOMETABOLISM

The capacity for culture, together with human dexterity, led to one particularly important difference between the ecology of our species and that of other mammals. The regular use of fire and the manufacture and use of tools added an extra dimension to the metabolism of human populations – referred to as *technometabolism*.

Technometabolism is defined as the pattern of flow of energy and materials into, through and out of a human population that results from technological processes. It contrasts with biometabolism, which is the flow of energy and materials into, through and out of human organisms themselves. Of course, some other animals use tools, but technometabolism on the scale seen in human populations is a new phenomenon in the history of life on Earth. It is of tremendous significance ecologically, and in many other ways.

In particular, the use of fire was a development of enormous ecological significance. It was the first example of the regular and deliberate use by humans of extrasomatic energy – energy, that is, which is used outside the human body, as distinct from the somatic energy which is consumed in food, flows through the human body and is dispersed in the form of heat.

It has been estimated that the introduction of the regular use of fire in human populations approximately doubled the per capita energy use, bringing the average total energy used per day per person (men, women and children) to about 14 MJ: that is, roughly 7 MJ used in biometabolism and 7 MJ in the burning of wood.

In the early farming phase of human history, and in the early urban phase, new technologies were introduced that resulted in some intensification of technometabolism. In particular, there was an input of various metals, especially iron and the combustion of wood as a source of energy for smelting. There has been an explosive increase in the intensity of technometabolism in Ecological Phase 4 of human history (the Anthropocene).

The analysis of flows of materials and energy into, through and out of urban systems has now become an important field in human ecology. An early example is the study of the metabolism of Hong Kong[16].

---

16 Newcombe, K, J. D. Kalma and A. Aston. 1978. *The metabolism of a city: the case of Hong Kong*. Ambio. Vol. 7, No. 1, pp. 3-15, and Boyden, S., S. Millar, K. Newcombe, B. O'Neill, 1981.

Patterns of urban metabolism have an important influence on the health of human populations and of the ecosystems of the biosphere.

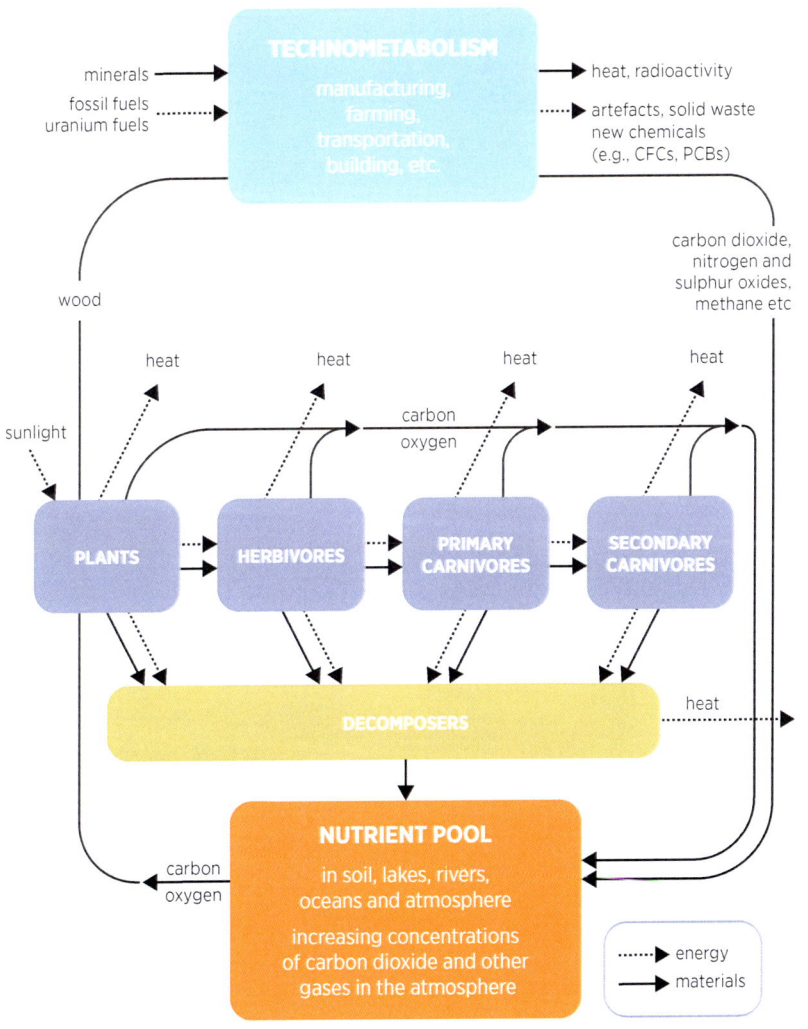

Figure 9: Flows of materials and energy in the modern world

TECHNOADDICTION

In the history of civilisation, it has often been the case that new techniques have been introduced simply for curiosity, or sometimes because they have benefitted a particular individual or group in society. But with the passing of time, societies have organised themselves around the new technologies, and their populations have become progressively more dependent on them for the satisfaction of basic needs. Eventually a state of complete dependence is reached.

The current dependence of human populations on fossil fuels is an obvious and extremely serious example. Other recent examples include our dependence on electricity and on computer technology. This insidious form of addiction passes largely unnoticed, although it is often of immense economic and ecological significance.

From the ecological standpoint it is significant that, in the modern cultural setting, the following basic behaviours often use up much more energy and create much more pollution than they did in the past: seeking in-group approval; seeking to conform; seeking attention; seeking novelty; seeking excitement; seeking variety; seeking comfort; visiting relatives; being selfish; being greedy; and being generous.

A GENERATIONAL PERSPECTIVE

Picture yourself on the stage of a large theatre with room for an audience of two thousand.

In your mind's eye, place your mother, as a young woman, in the seat at one end of the front row, and then her mother next to her and so on – until you have filled the theatre with 2,000 generations of young mothers and daughters.

When I perform this exercise, only my mother and the two mothers beside her would have seen motor cars. Only the women in the front six or seven rows would have lived after the earliest cities came into existence, although few of them are likely to have lived in cities; and only those in the front twenty or so rows would have been alive since farming began.

All the remaining 1,600 or so women would have been hunter-gatherers. You could fill five more similar theatres with earlier maternal hunter-gatherer ancestors belonging to the species *Homo sapiens*. All these women really existed, and they were all healthy enough to produce a living daughter.

If you were to carry out the same mental exercise imagining you were in a stadium with seats for 150,000 people, the ladies in the rows at the back would be australopithecines.

HUMAN RIGHTS & THE RIGHTS OF NATURE

The notion that humans possess 'rights' is ancient. It is inferred 4100 years ago, in the Code of the Sumerian king, Ur-Nammu, and 300 years later in the Code of the Babylonian king, Hammurabi. However, the actual expression 'human rights' only came into common parlance after the General Assembly of the United Nations adopted and proclaimed the Universal Declaration of Human Rights in 1948.

The natural sciences have nothing to say about human rights. Nature does not bestow rights on us. We can decide to confer rights on ourselves and others, and acceptance of these rights can then become embedded in our culture. The concept of human rights is thus a cultural construct. This does not mean, of course, that it is unimportant.

In recent years there has also been much discussion about the notion of animal rights. Indeed, some authors believe the concept should be extended not only to animals, but also to trees, mountains, rocks and the streams.[17]

The Australian philosopher Peter Singer has been a prominent supporter of animal rights, although he does not extend the concept to trees, mountains and rocks which, he says, do not feel anything, and so do not have rights.

As in the case of human rights, animals do not possess rights until they are conferred on them by humans. Animal rights, like human rights, are a product and component of human culture.

---

17   I have not followed this literature, but an excellent account of this movement in the English-speaking world is provided in the book *The rights of nature: a history of environmental ethics* by Roderick Nash.

PESSIMISM & OPTIMISM

So, after 4000 million years of biological evolution on our beautiful planet, we have an animal with a big brain that calls itself *sapiens*, applying its intelligence to activities that could well bring an end to the species to which it belongs, as well as countless other forms of life. There are two areas of major concern, namely ecological threats and military threats. Both are on a massive scale, and there is overlap between them.

It is apparent that the prevailing cultures across the world today condone, if not encourage, the activities that lie behind these ecological and military threats to our future. It follows that there is no hope of ecological survival, or long-term peace between nations, unless there come about radical changes in the worldviews and priorities of these cultures.

What are the chances of cultural enlightenment coming about in time to prevent global catastrophes on a massive scale?

Regarding the ecological issue, I am rather pessimistic.
The maladaptive assumptions of the prevailing cultures are deeply ingrained, and it seems unlikely that effective cultural reform will happen soon enough to avert ecological collapse. However, I am optimistic enough to believe this is not impossible; and so long as this is the case, then surely those of us who understand the nature and severity of the current situation should be doing all we possibly can to bring about this radical cultural transformation.

Similarly with warfare, I am not optimistic. But again, enlightened cultural reform is not impossible. Imagine a world in which warfare was unthinkable! Again, I believe that the biohistorical perspective can make an important contribution. It certainly highlights the gross insanity and immorality of war.

I have heard it argued that there is no hope of achieving ecological sustainability until we have a much better understanding of the biophysical and social system in which we live, and that much greater effort should be aimed at achieving such understanding through systems modelling. In my view, despite recent advances in systems theory and information technology, the complexity of the system is such that this kind of understanding will always be beyond us.

However, all is not lost. I suggest we do not need to understand all the intricacies of this massive and extremely complicated system in order to move forward to a biosensitive society. All that is required, initially, is a single change in the system – the insertion of biounderstanding into all the prevailing cultures worldwide.

This single cultural change would have far-reaching repercussions throughout the whole of society – the butterfly effect of chaos theory. It would lead, first, to changes in the worldviews and priorities of the prevailing cultures themselves. Unlike the situation today, these cultures would hold profound respect for nature, and the achievement of harmony with the processes of life would be given the highest priority in human affairs.

This fundamental shift in worldviews would be followed by the introduction of new biosensitive cultural arrangements (e.g. economic systems, government regulations, population policies) leading to biosensitive human activities (e.g. energy use, food production, forestation, manufacturing, consumer behaviour, lifestyles).

Naïve? Unrealistic? Perhaps; but if so, then I think there is little hope for humanity.

Blue Mountains Upland Swamp (Garguree, Katoomba), a special Aboriginal place where a nature rehabilitation program is underway. (Photo courtesy: Sally Tsoutas, WSU)

PART 4:
PERSONAL PERSPECTIVES

The richness I achieve comes from Nature, the source of my inspiration. I have no other wish than to mingle more closely with Nature and I aspire to no other destiny than to work and live in harmony with her laws.[18]

Claude Monet

Nature is my god. To me, Nature is sacred. Trees are my temples and forests are my cathedrals.[19]

Mikhael Gorbachev

Ethics is nothing other than Reverence for Life. Reverence for Life affords me my fundamental principle of morality, namely, good consists in maintaining, assisting and enhancing life, and to destroy, to harm or hinder life is evil.[20]

Albert Schweitzer

---

18 www.wisdomquotes.com/quote/claude-monet.html
19 "Nature Is my God" - interview with Fred Matser in *Resurgence* No. 184 (September-October 1997). Pg 14-15.
20 Civilisation and Ethics, 1923

Nature

All religions have their stories. Christians, Muslims, Sikhs, Buddhists, Hindus and Judaists – they all have their stories. All political ideologies have their hallowed texts. I have my story. It is the story of life on this planet. It is an amazing story, and full of meaning.

The more I learn about life on Earth, through personal observations and from science, the greater is my sense of wonder, and the more profound my sense of respect, indeed feeling of reverence, for nature and for the creative processes that gave rise to the living world.

I am thinking not only of the natural environment, with all its mindboggling diversity and beauty, but also of the amazing and extremely complicated processes that go on inside my own body, and that have kept it going for over ninety-eight years.

Four billion years ago, the crust of the Earth consisted of a mass of minerals, an atmosphere and some energy. It also received a constant supply of energy from the sun.

It turned out that this primordial system was endowed with amazing properties. Left to its own devices, it eventually gave rise to living organisms, starting with single-celled microbes, but leading to the myriad of life forms that exist on our planet today, and to human civilisation.

It is said that life is ultimately explainable in terms of the laws of physics. These laws will lead, willy-nilly, to the creation of life on suitable planets. I am not competent to pass judgement on this hypothesis; but let us suppose it is correct – then we are still left with the question 'Where did the laws of physics come from?'

And indeed, what a wondrous set of laws – laws that have led to the eventual coming into being, starting from the primordial mass of matter and energy, of the vast array of living organisms that live on our planet today, as well as the plays of Shakespeare and the music of Mozart.

Thus, science still leaves an unexplained mystery – the mystery of the origin of the natural phenomena and laws that gave rise to the living world. True, it is postulated that the universe started with a big bang; but this does not explain how the big bang came about, nor where all the matter and energy that dispersed in the big bang came from.

The underlying mystery of existence is responsible for me, and for all life on Earth. I believe it will always remain a mystery.

Evolution

Darwinian theory and modern genetics have provided an explanation of the evolution of life on Earth. However, I am among those who, while accepting the role of natural, sexual and artificial selection in evolution, wonder whether there might not be some other, as yet unknown, influence.

For instance, in my view sexual selection does not adequately explain the eyespots on the peacock's tail, and the arrangement of colours on the hundreds of individual barbs that make up the eyespots.

The eyespot on the peacock's tail feather (Photo: Stephen Boyden)

Sexual selection has been held responsible for the wonderful range of colours and shapes in the plumage of the numerous species of Birds of Paradise in Papua New Guinea and of Manakins in the tropical forests of America. Is it responsible for the amazing range of colour patterns seen in butterflies and moths?

Another group of animals that exhibit an extraordinary range of colours and shapes is the Nudibranchs, the soft-bodied marine molluscs that shed their shells when they become adults. There are over 2,000 species of nudibranchs. What is the evolutionary explanation of this spectacular array? Sexual selection?

Similarly, I cannot imagine how natural or sexual selection could possibly bring about the beautiful sand circles created by the male, white-spotted puffer fish.

The male Puffer Fish's pattern in the sand (Photo: Kawase et al., 2013)

There is a fantastic range of different shapes, sizes and colours among flowers. Is this simply due, as evolutionary botanists maintain, to selection pressures arising from interaction with pollinating insects and birds?

Turning to our own species, I find it difficult to picture how natural selection in the evolutionary environment could have given rise to some human characteristics. What kind of selection pressures were responsible for a concert pianist's ability to memorise the sequence of thousands of musical notes, and combinations of notes, in a Beethoven piano concerto, and then their ability to play these notes, through extremely rapid and complicated movements of their ten digits, on the piano keyboard; and their ability to recall all the thousands of notes in dozens of other pieces of music?

So, for me, there is a mystery; and I do not rule out the possibility that there is some other kind of influence on gene selection. I have no idea what it might be.

Ethics

While science tells us about the evolution of life on Earth and about the ecological and physiological processes that keep us alive; it has nothing to say about morality. It does not tell us whether it is right or wrong to cause pain or distress needlessly in humans or other animals. It does not tell us whether it matters if humankind trashes the living systems on which it depends, or if our species comes to an early end because of its own activities and arrogance.

As I see it, we have a choice. We can seek to live in harmony with nature, or we can look on nature with disdain, and set out to exploit and conquer it. For me, there is no doubt whatsoever. My understanding of the story of life and of our biological origins leads me to choose the first of these pathways. I believe it is not only wise, in terms of my survival and wellbeing, but also morally right to strive to live in harmony with nature. I see this as of supreme importance.

I consider it morally wrong, the height of disrespect and arrogance, to desecrate the living systems of our beautiful planet. I have no sympathy with the notion, dating back to the so-called Enlightenment, that we should aim to conquer nature – to conquer the living system that gave rise to us, of which we are a part, and on which we are totally dependent.

So my understanding of life on Earth, and how we all came to be here, leads me to feel morally committed to strive for harmony with nature, and with other humans as part of nature.

Spirituality

I should also say that, apart from the sense of wonder and joy that I get from observing and learning about life on Earth, experiencing nature also has meaning for me that can best be described as transcendental, or spiritual. When I am alone in the natural environment, I sometimes suddenly feel not only a deep sense of awe, and reverence, but also an apparent awareness of another dimension of reality – hard to describe, but very real. The feeling is heartening, strengthening, calming, comforting and inspiring. This happens especially when I am in the wilderness, but I increasingly experience it in other situations, like in the room in which I now live.

What is the explanation of these experiences? I do not know, but they are an important part of my life experience. I suppose many, if not most, people have them. Jane Goodall, for example, says:

> *"All the time I was getting closer to animals and nature, and as a result, closer to myself and more in tune with the spiritual power that I felt all around. For those who have experienced the joy of being alone with nature there is really little need to say more; for those who have not, no words of mine can ever describe the powerful, almost mystical knowledge of beauty and eternity that come, suddenly and all unexpected".* [21]

The story of life reminds me that I am not only a product, but also an intrinsic part of nature – even if an infinitesimally small part. I am as much a part of nature as are the eucalypts on Oakey Hill, the cockatoos in the garden and the wallaroos at the farm. I find myself feeling at one with the living world. It is a good feeling. It gives me peace of mind.

---

21   https://www.goodreads.com/author/quotes/18163.Jane_Goodall

Roberts Creek, Tinderry, NSW (Photo: Karina Bontes Forward)

SOME PUBLICATIONS

Boyden, S. (Ed.) 1970. *The impact of civilisation on the biology of man.* ANU Press. Canberra.

Boyden, S. 1979. *An integrative ecological approach to the study of human settlements.* MAB Technical Note No.12. UNESCO, Paris.

Boyden, S., S. Millar, K. Newcombe, B. O'Neill, 1981. *The ecology of a city and its people: the case of Hong Kong.* ANU Press, Canberra.

Boyden, S. 1987. *Western civilization in biological perspective: patterns in biohistory.* Oxford University Press, Oxford.

Boyden, S., S. Dovers, M. Shirlow. 1990. *Our biosphere under threat: ecological realities and Australia's opportunities.* Oxford University Press, Melbourne.

Boyden, S. 1992. *Biohistory: the interplay between human society and the biosphere - past and present.* Parthenon/UNESCO, Paris.

Boyden, S. 2004. *The biology of civilisation: understanding human culture as a force in nature.* UNSW Press. Sydney.

Boyden, S. 2005. *People and nature: the big picture.* Nature and Society Forum. Canberra.

Boyden, S. 2011. *Our place in nature: past, present and future.* Nature and Society Forum. Canberra.

Boyden, S. 2013. Human biohistory, Chapter in S. Singh, H. Haberl, M. Chertow, M. Mirtl, M. Schmid (Eds.), *Long term socio-ecological research: Studies in society-nature interactions across spatial and temporal scales.* Springer. Netherlands.

Boyden, S. 2016. *The bionarrative: the story of life and hope for the future.* ANU Press, Canberra.

Boyden, S. 2016. *The biohistorical paradigm: the early days of human ecology at the Australian National University.* Human Ecology Review. 22. No1. Pp. 25-47.

Boyden, S. 2022. Phylogenetic maladjustment and the biology of modern society. Chapter 1 in *Spiritual motivation.* Vol.2. *New thinking for a post-Covid world.* Edited by Jeremy Ramsden. Collegium Basilea. Basel. Switzerland.

FURTHER READING

Bill Gates. *How to Avoid Climate Disaster.* Penguin.

Greta Thunberg. 2021. *No One is too small to make a difference.* Penguin Classics.

Paul Hawken (Ed.). 2017; *Drawdown: The Most Comprehensive Plan Ever Proposed to Reverse Global Warming.* Penguin.

Paul Collins. 2021. *The Depopulation Imperative: How Many People Can the World Support?* Australian Scholarly Publishing.

https://ourworldindata.org/world-population-growth

Herman Daly. 1997. *Beyond Growth.* Beacon Press.

Tim Jackson. 2009. Prosperity without Growth: Foundations for the Economy of Tomorrow. Routledge.

Robert Constanza and Ida Kubiszewski. 2014. *Creating a Sustainable and Desirable Future: Insights from 45 Thought Leaders.* World Scientific.

Jeff Olson. 2012. *The Third Mode: Towards a Green Society.* Jeff Olson.

Naomi Klein. 2015. *This Changes Everything: Capitalism vs the Climate.* Penguin.

Charles Massey. 2020. *Call of the Reed Warbler: A New Agriculture – A New Earth.* University of Queensland Press.

George Monbiot. 2022. *Regenesis: Feeding the World Without Devouring the Planet.* Penguin.

Julian Cribb. 2016. *Surviving the 21st Century: Humanity's Ten Greatest Challenges and How to Overcome Them.* Springer.

Thomas Berry. 1999. *The Great Work: Our Way to the Future.* Crown.

Robert Dyball and Barry Newell. 2023. *Understanding Human Ecology: A System Approach to Sustainability.* Routledge. 2nd edition.

Stephen Boyden graduated in Veterinary Science in London in 1947. From 1948 to 1965 he carried out research in immunology in Cambridge, New York, Paris, Copenhagen and Canberra.

From 1965 until his formal retirement at the end of 1990 he pioneered work on human ecology and biohistory at the Australian National University (ANU). In the 1970s he initiated and was Director of the Hong Kong Human Ecology Program – the first comprehensive attempt to study the ecology of a city. For some years after that he was a consultant for UNESCO on urban ecology. In 1973 he introduced undergraduate courses at ANU in Human Ecology and these have survived to the present day. He has continued to work in this field since he retired.

He has published nine books on human ecology and biohistory, five of them since his retirement. He is at present a Professor Emeritus at the Fenner School of Environment and Society at ANU.

Candlebark, *Eucalyptus rubida* (Photo: Stephen Boyden)

www.ingramcontent.com/pod-product-compliance
Lightning Source LLC
Chambersburg PA
CBRC101141030426
42334CB00012B/126